# REPTILES

Rebecca Woodbury, Ph.D., M.Ed.

Gravitas Publications Inc.

# REPTILES

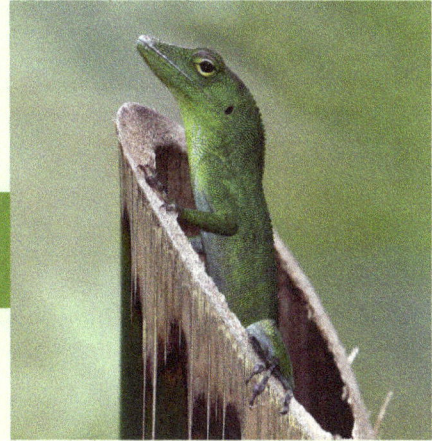

Illustrations: Janet Moneymaker

Reptiles
ISBN 978-1-950415-67-0

Published by Gravitas Publications Inc.
Imprint: Real Science-4-Kids
www.gravitaspublications.com
www.realscience4kids.com

RS4K

Photo credits: Cover Title Pg.: By LMPark Photos, AdobeStock; Above: Photo by VD Photography on Unsplash; P.3. ivabalk from Pixabay; P.5. Nikki from Pixabay; P.7. Top, Balaji Malliswamy on Unsplash; Bottom, David Cashbaugh on Unsplash; P.9. Mitch Gritts on Unsplash; P.11: 1. Flicka, CC BY SA 3.0; 2. Benny Trapp, CC BY SA 3.0; 3. Monika from Pixabay; 4. Dominik Weigelt on Unsplash; P. 13. David Clode on Unsplash; P.15. Alexis Montero on Unsplash; P.17. Public Domain; P.19. Brocken Inaglory, CC BY SA 4.0; P.19.: 1. Chris Curry on Unsplash; 2. Room237, CC BY S 3.0; 3. Greg Schechter, CC BY SA 2.0; 4. Jack Kelly on Unsplash

Have you ever seen a
**turtle** snapping at a leaf?

I've seen them eat watermelon and bugs too.

Have you ever seen a **snake**
slithering through the grass?

Do you live near **crocodiles** or **alligators** with their big toothy jaws?

Do you think they brush all those teeth?

Probably not.

Maybe you have seen a striped **lizard** speed across the ground in front of you.

Lizards run fast!

Turtles, snakes, crocodiles, alligators, and lizards are all **reptiles**.

Reptiles are a group of animals that share some things in common.

There are lots and lots of different kinds of reptiles!

Reptiles have skin that is covered with **scales**.

What a pretty lizard!

Reptiles keep their body at
the right temperature by lying
in the sun to get warm. They
lie in shade to cool off.

Most reptiles lay **eggs** that
have soft leathery shells.
Other reptiles give birth to
live babies.

Look! Those eggs are starting to hatch.

I wonder what kind of reptile will come out?

Many reptiles live on land.

Some reptiles, like crocodiles and alligators, live much of their life in water.

Some reptiles, like the sea turtle, spend all their time in the ocean and only come out on land to lay eggs.

It is fascinating to study the many different types of reptiles.

What reptiles have you seen? What have you observed about them?

1

2

3

4

# How to say science words

**alligator**   (AA-luh-gay-tuhr)

**crocodile**   (KRAH-kuh-diyl)

**egg**   (AIG)

**lizard**   (LI-zuhrd)

**reptile**   (REP-tiyl)

**scale**   (SKAYL)

**science**   (SIY-uhns)

**snake**   (SNAYK)

**turtle**   (TUHR-tuhl)